Chat to your Cat

Lessons in Cat Conversation

By Martina Braun

Im Dorfe 11, 22946 Brunsbek, Germany

Translation: Andrea Höfling
Cover design and layout: Ravenstein + Partner
Cover and all other photos: Fotonatur.de, Urs Preisig, Ulrike Schanz
Editorial: Anneke Bosse and Christopher Long
Printed by: Westermann Druck, Zwickau

British Library Cataloguing in Publication Data
A catalogue record of this book is available from the British Library.

Printed in Germany
ISBN 978-3-86127-966-2
www.cadmos.co.uk

Contents

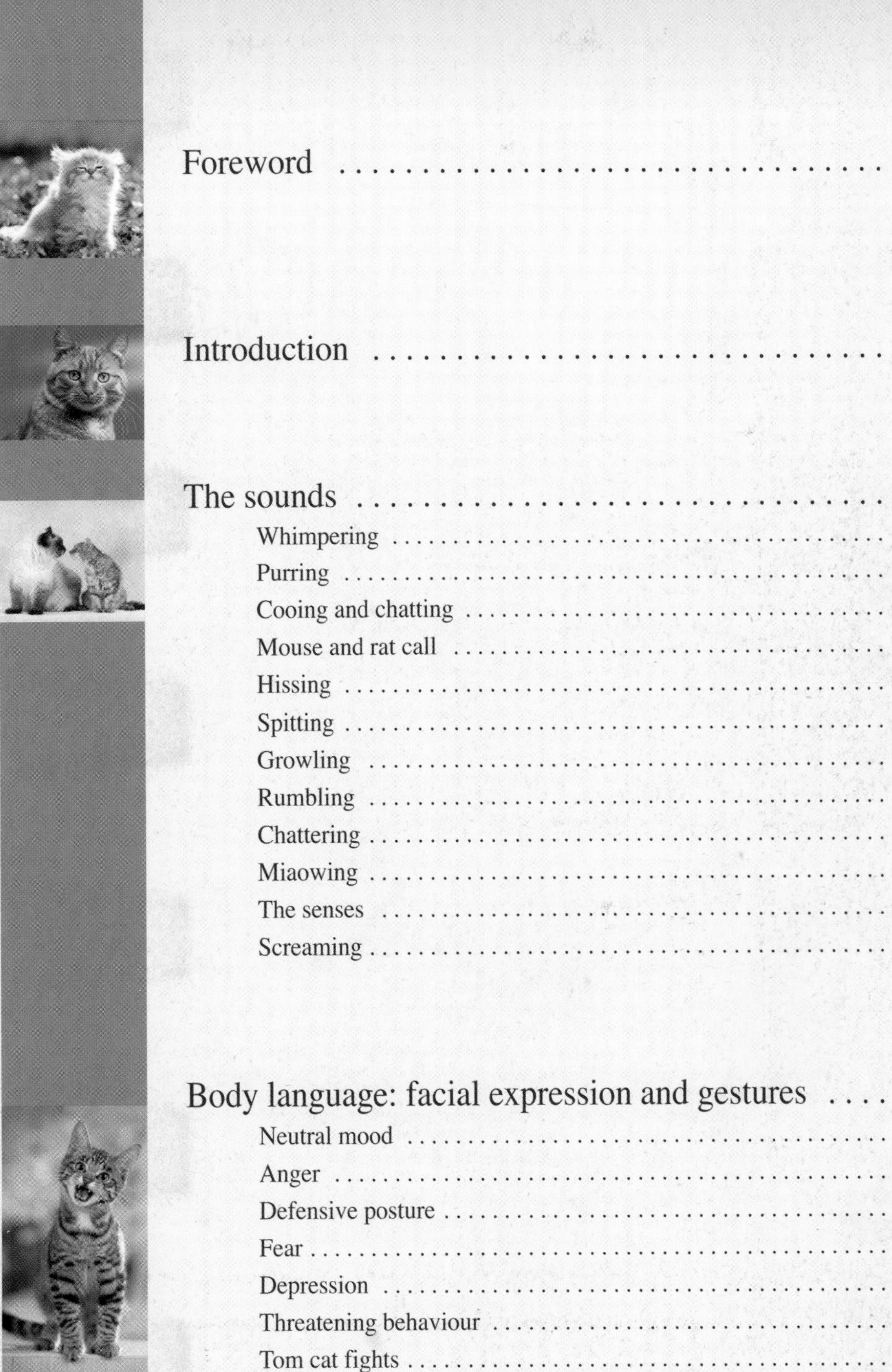

Contents

Foreword

With her book Chat to your Cat – Lessons in Cat Conversation, Martina Braun has made an informative and entertaining contribution to further our understanding of cats, and also the understanding between cats and humans. In order to build and sustain a harmonious relationship – between two species or within the same species – every potential cat owner has to develop an understanding of the behaviour and the nature of these predators, who share their lives with us to a large degree of their own free will. This understanding is a prerequisite for dealing with these pets in a responsible, trustworthy manner. Equally, ethologists and animal psychologists have the responsibility to further the dissemination among cat owners of science-based new findings – and ensure a layperson can understand them. This, too, Martina Braun has succeeded in doing in her book.

Dr Dennis C. Turner PhD
Director of the Institute of Practical Ethology and Animal Psychology (Institut für angewandte Ethologie und Tierpsychologie, I.E.T.), Hirzel/Schwitzerland.

(Photo: Preisig)

Hello. My name is Sala!

Introduction

May I introduce myself? I am Sala, the recurring thread on four paws who will accompany you throughout this book. I'm sure my human co-author will do her best to acquaint you with cat behaviour and our other idiosyncrasies. However, she's only human, and not a cat, after all! Therefore it is absolutely essential that I supervise the whole project in my capacity as a genuine tom cat to ensure that things aren't lost in translation. There have already been so many misunderstandings throughout the history of my species!

A long time ago (in about 2600 BC), for a period of 1,300 years, humans thought of we cats as divine beings. From the cats' perspective it was purely a symbiotic relationship from which both sides were able to benefit. In no way had we ever given up our freedom and independence! We cats kept the grain stores free from those pesky rodents that were causing a lot of damage. The humans kept their grain, while we cats developed comfortably plump tummies due to the rich pickings of rats and mice. Bastet, the goddess of cats, was regarded as the goddess of fertility, joy, dance and festivities, as well as the protectress of pregnant women. As a result, cats, too, came to be adored and revered as divine beings. But this privilege came at a very high price. The priests of Bastet bred and then sold us to all and sundry. Destined to serve as sacrifices, our smaller brethren had their necks wrung, while larger and stronger specimens had their skulls smashed. Then, their bodies were mummified and sacrificed to Bastet. Many years later, thousands of my fellow cats were found in the Valley of the Kings in a mummified state. I ask you: was this really necessary? Of course, we are divine beings! Who could ever doubt that?! But honestly, that's no reason to kill and mummify anyone!

As humans became explorers and traders, we cats too conquered the whole world, and this is how we arrived in Europe. My ancestors were in raptures! But human stupidity soon caught up with us. When Christianity established itself as a religion between the 11th and 14th Centuries, we were unfortunate enough to be caught up in its machinations.

Even before that time, there was an old Celtic belief that cats had once been human, and had been changed into felines to punish them for their wicked ways. Later, Catholic culture latched on to those old pagan superstitions, and the cat became thought of as the witch's familiar.

The Inquisition came down upon cats and witches with a vengeance, and from the 13th Century onwards, horrible atrocities were committed against felines and humans alike. Black cats in particular were believed to be agents of the devil, especially if owned by elderly women.

Cats and witches have had a long association with Hallowe'en. Because we cats are nocturnal creatures, we came to be seen as the evil servants of the witches, out to do their bidding under cover of darkness. Some humans even believed that witches had the power to change into cats in order to carry out their wicked deeds more easily and escape detection.

Because we cats were accused of being in cahoots with Satan and witchcraft, my ancestors were shunned, and right up to the middle of the 18th Century, many tens of thousands of us were burned all over Europe. Honestly, I ask you: who was the villain of this deplorable episode in history? To me, it looks suspiciously like whoever it was, it wasn't the cat.

Fortunately, nowadays we cats are well-loved as pets, but we still evoke overwhelming feelings of either love or hatred in humans. Many humans still aren't able to interpret our behaviour correctly.

So in order that we may understand each other even better in future, this book is going to attempt to teach the reader a little bit of our cat 'lingo'. I will teach you humans yet! Enjoy!

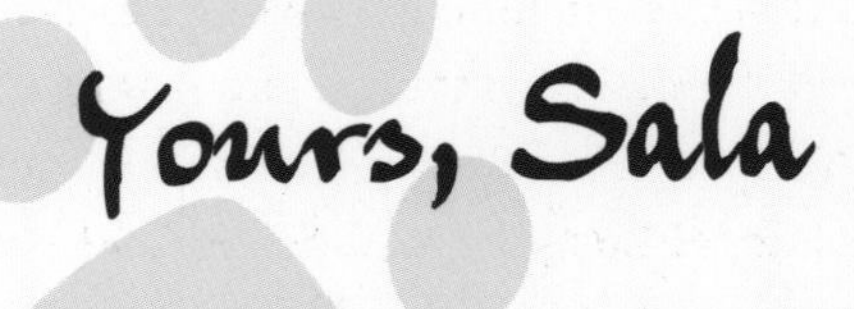

(Photo: Schanz)

The sounds

When thinking about communication between living beings, that in the form of sound often springs to mind first. The term 'communication' (Latin, communicare: sharing, conveying, participating, acting jointly, uniting) is the comprehensive term for a mutual exchange of thoughts and feelings, not just by using sound (acoustic), but also via body language, gestures and facial expression (visual) and depositing odours (olfactory).

When individuals communicate, they relate to each other. The reaction of one depends very much on the actions of the other, and vice versa. This highlights the importance of communication in establishing common ground and avoiding serious conflicts. The tools available to the cat for this purpose are manifold. Bearing in mind what effective hunters cats are, any misunderstanding can create the risks of serious injury to both parties. In fact, it is the large number of subtle nuances and facets the cat has at his disposal that make it so difficult for we humans to interpret them correctly. Let's start with the one aspect of language through which humans understand each other best: communication via sound.

In the past, many efforts have been made to count and classify the individual sounds that cats produce. We now distinguish between six basic sounds: purring, miaowing, level one hissing, level two hissing, screaming and growling. Further scientific research has concluded that domesticated cats have 16 different sounds at their disposal, and has categorised them into three different groups:

- Murmuring (sounds made while the mouth is closed);
- Vocal sounds (for communication purposes with the cat's human, produced while the mouth is being gradually closed); and
- Sounds of high intensity (sounds produced with an open mouth, while the size of the mouth opening varies). This is mainly reserved for communication with other cats.

The different sounds are not always easily distinguishable. When a cat is cornered by a human or by a pushy cat, he may start showing his displeasure by sounding an irritated miaow. If that's not enough, this miaow may seamlessly merge into a hissing or growling sound, and if the tormentor still doesn't back off, the whole thing may escalate into an angry rumbling. The type as well as the intensity of the sound varies according to the situation, and the transition from one sound to another is flowing. As a result, any attempt to list these sounds, including the one that follows here, can only be a rough outline of the acoustic communication signals of which a cat is capable.

Whimpering

The first sound that a kitten makes is whimpering, which serves to trigger the mother's maternal affections and care. The feline behaviourist Paul Leyhausen has proved that the maternal action of carrying kittens back to the nest can only be triggered by the kitten's whimpering sounds. If a kitten has fallen out of the nest and is crawling about, but doesn't whimper, the mother takes no action. Only when the kitten sounds his 'meeeh' will the mother carry him back. Why? Well, bringing up kittens is a strenuous activity, for both parties. In order not to needlessly waste energy and effort, a clear marker or trigger is necessary. The whimpering puts a lot of stress on the little lungs, so this guarantees that a kitten only whimpers when it is absolutely necessary. The mother's behaviour is adapted with according efficiency. She will only carry him back to the nest when he cries for help.

Like purring, these initial, very early sounds belong with the group of 'vocal touch sounds' and are primarily designed to build and reinforce social attachments. You could say the animals use a 'vocal' touch either when the physical touch is absent, or in order to reinforce it.

Purring

The little tyke has to attract mums attention by whimpering for her to carry him back to the nest. (Photo: Fotonatur.de/Askani)

The kittens' first purring sounds can be heard, albeit very quietly, almost straight after birth when they suckle from their mother. A kitten is able to swallow, suckle and purr all at the same time. By using this particular vocal touch sound, he conveys to his mother a sense of well-being. This way she knows that the little one is well without having to get up and thereby perhaps interrupting the feeding process. The purring is answered. The mother also purrs while she is nursing her young. In doing so, she soothes her offspring, as well as herself.

All cat-like animals (felines) have the ability to purr, not just domestic cats. Adult cats living in the wild purr almost exclusively when they have young. The domestication of the cat basically led to a permanent state of adolescence. As a result of living with humans, our domestic cats have kept their ability to purr, and to signal well-being, into adulthood.

Purring is a vibrating sound at a low frequency between 27 and 44 hertz. Cats also purr when they are in pain, and when they are sick or dying. Therefore it is assumed that they have the ability to calm themselves by purring. Adolescent cats that are playing with other adult cats sometimes purr in order to emphasise the peaceful nature of their playing towards the superior playmate, and to calm themselves. Only extremely anxious and frightened cats, or cats in an extremely aggressive mood, don't purr.

By the way, there is a good reason why kittens are born with hair, but blind and deaf. If they were already able at this point to perceive all the stimuli of their environment with their eyes and ears, they'd be frightened and confused, maybe curious, but definitely distracted from the main issue: suckling. Their life would be over in a matter of hours.

At birth, the only functioning senses that healthy kittens have are touch and smell. In order to prevent the kitten from getting too far away from the nest and wasting vital energy during his search for mum and the protection, warmth and milk she offers, he will crawl on his tummy in small circles, usually in an anti-clockwise direction. He realises he has found her when he can feel her warmth (tactile) and smell her milk (olfactory). By swaying his head gently from side to side, the kitten searches the skin surface of his mother's tummy (sense of touch) in order to locate the prominent teats that he takes into his mouth to suckle. At the same time, the tiny paws left and right of the teat begin to tread and massage rhythmically. The treading action stimulates the flow of milk. This infantile instinctual action survives into the cat's adulthood when human and cat are living together. The cat will use it when jumping on a soft surface, or when jumping on to your lap for a cuddle, by stomping around with his front paws, before happily settling down.

Even if the claws sometimes sting a little, this is a token of the cat's trust in you, and it would be wrong to chase him off or punish him because of it.

Sadly, some cats, such as myself, have not had the good fortune of growing up in close contact with humans, having found proper homes only later in life.

Although we love them dearly, and all those peculiar things they get up to, that's why cats like me sometimes still feel a little insecure when dealing with humans.

Do you want to know my own recipe for success? It's simple: **I just purr!** You know, those tense moments like being picked up, or visiting the vet's and so forth; purr, purr, purr – it works a treat! Humans tend to be absolutely delighted, and give me lots of tender love and care. At the same time, it also soothes my own nerves and helps me keep my cool. It's a marvellous thing, purring.

Cooing and chatting

All cooing sounds serve as a friendly greeting between familiar individuals, either between cats or between cat and human. If the cooing is accompanied by a quiet miaowing, this is also called chatting.

As early as 10 days after the birth, the mother cat begins to coo upon returning to the nest, and it can therefore be assumed that this sound

Cooing is a friendly sound that the cat also uses to greet his trusted and familiar human. (Photo: Schanz)

may also be categorised as a social 'vocal' touch. The mother stays by the side of the nest and begins to coo persistently until her kittens wake up and start their characteristic whimpering. After a few days, the mother will expect the kittens to come crawling towards her in response to her persistent cooing.

Sometimes you can overhear cats that are very close and familiar with each other having a 'chat' using variations of this cooing, which are unique to these two individuals. This is called dyadic dialect (Greek, dyade: twosome). Similarly, you, being a trusted and familiar human being, as well as an esteemed tin-opener, will also be greeted by your moggy with a friendly cooing sound.

Mouse and rat call

The mouse call that the mother uses upon bringing her kittens their first mice at the age of four to six weeks is a slightly more guttural version of the cooing sound. With this adapted cooing sound, she signals to her kittens that she has brought them something exciting, and beckons them to come closer to see what it is.

My own 'big game huntress' Anima will announce from afar that she has been successful, and that she is approaching the house with a mouse. For well-practised owners of outdoor cats, this is the signal to quickly shut the doors to all bedrooms and living rooms! Unfortunately, this call does not give you any clue whether this mouse is already dead or whether it will be darting around your flat in a very lively manner indeed!

We owe one further amazing observation to Paul Leyhausen. He delivered the proof that cats really do have the ability to use linguistic terms. Previously, this ability had only been attributed to primates and, of course, humans. He overheard mother cats that brought their kittens a rat – or perhaps even some rat portions, which may be much smaller than a whole mouse – utter a shrill, often drawn out scream that was quite different to the cooing mouse call. This is called the rat call.

Even to a fully grown cat, a rat is a dangerous prey, able to defend itself, and the kittens display a prompt reaction. When they hear the mouse call, they will approach without hesitation and show a keen interest. The rat call on the other hand brings about clear signs of distrust and caution in the kittens, who will only approach slowly and with a crouched body posture. The kittens obviously understand the difference between the two calls without having experienced the meaning for themselves.

Hissing

The hissing sound also develops early, although very young cats make it without blowing out air. They open their little mouths about halfway and make the facial expression that traditionally accompanies the hiss. The perfected version of this will later look like this: the upper lip is lifted and the tongue, particularly at the sides, is curved upwards almost to the roof of the mouth. This enables the cat to breathe out sharply, which produces the typical hissing sound.

Close up, this expulsion of air can even be felt. This is the reason why cats find it unpleasant and off-putting to have someone blow into their face, a fact we can use for the purpose of upbringing. But please, only use this in order to fend off excessively rough physical behaviour that is getting out of control, when your cat is using his claws during play, for example.

Those humans, I really don't get it sometimes! When they happen to encounter one of us cats outside, some humans will approach us, all kindness and smiles, clearly wanting us to come over, so they can stroke us. But you won't believe what they do to next! They start making hissing noises, something like: 'Bssss bsss bssss'.
Don't they realise how off-putting this sounds to a cat? If you were to hear a snake hiss, would you rush over to say hello? I think not!

The sharp expulsion of air produces the typical hissing sound that is warning to an adversary in addition to the facial expression. (Photo: Fotonatur.de/Meyer)

Spitting is a warning sound that is meant for non-feline adversaries and is often accompanied by the typical arched back. (Photo: Schanz)

Used too often and incorrectly this disciplinary measure can make the cat scared of humans who appear to him to be like gigantic cats.

The more emphatic the warning, the more seriously it is taken. With this voiceless sound, the cat threatens his adversary not only visually, (facial hissing expression), but also by touch (blowing air), as well as acoustically (hissing sound). It is the last chance to avert a hefty whack with a paw.

Spitting

The spitting sound is produced by the expulsion of air in sharp and explosive blows. It is a warning sound, and its exclusive purpose is to impress and baffle a non-feline adversary in order to gain time for a quick getaway, or to reach a more favourable position. The spitting is not used for communication between cats. It is often accompanied by an arched back.

As you can see, the correct arching of the back also has to be practised. (Photo: Fotonatur.de/Morsch)

Growling

We're all familiar with the menacing sound of a growl, not only from cats, but also from dogs and even rabbits! There are certain signals that are understood by all species – even humans. This type of communication is inter-species. It makes sense, because it is used when basic sentiments are involved, such as warning, defence and fear. As a secondary characteristic, it can often be observed that an animal will make itself appear larger by making its fur stand on end, walking on tip-toes and arching its back, to deter and deceive an adversary – for example, birds puff up their feathers, dogs and cats make their fur stand on end. If a cat feels seriously threatened or cornered, the hiss may turn into a growl. It is advisable to take a growling cat seriously, because this way he signals unambiguously that he will go on the attack, if necessary, and that he will also bite.

The depth and fervour with which even adolescent cats are able to growl is amazing. Growling is not necessarily reserved for emergencies. When cats play with each other they'll test the effectiveness of their growl on their playing partner – even if the disputed object is only a toy mouse or a mundane bluebottle.

The rumble is the last shot across the bow. (Photo: Fotonatur.de/Meyer)

Rumbling

Rumbling is usually used by adult cats as part of inter-feline communication, usually only when another cat becomes too pushy. It is a vocal increment of the growl and is used as a warning: 'That's enough! Stop, or there'll be trouble!'

Chattering

The meaning of the chatter has not yet been fully explored. Cats display this behaviour when they concentrate on a prey that they very much desire, but can't reach. The mouth is slightly open and the cat chatters, stutters or bleats. It is probably some kind of a displacement activity (see box on page 33).

By the way, a cat sitting behind a window pane chattering with dedication whilst observing a bird, or grumbling whilst observing another cat, may very well be in the process of bottling up a motivation that is not acted upon, and he should not be touched in this situation. The cat may redirect his aggression, turn around with the speed of lightning, and lash out at the hand that's cuddling him. A cat that

If a cat is concentrating on life on the other side of the window, he should not be touched suddenly, because he may be engaged in a virtual visual hunt and is likely to be very tense. (Photo: Fotonatur.de/Meyer)

is in the middle of such a 'virtual' hunt or duel (conducted only with the eyes) is enormously tense, and does not expect to be touched. This means strictly hands off; whoever gets a whack with a paw in this situation only has himself to blame.

Miaowing

Cats open their mouths wide in order to miaow, and then close it again slowly. This creates high pitched, vocal sounds that can vary a lot. There is no cat owner in the world who won't recognise the demanding miaowing when the food bowl is empty once more! Should the human not react as speedily as possible – at this point I find the cat's proverbial patience sadly absent – the whole tonal range is employed, from dirge-like moaning to nagging and ranting!

My British short-hair tom Mogli is a lovely, soothing exception, because he limits his begging attempts to an almost hummed 'Mmmmhhh', similar to nursery school children learning the alphabet. 'Mmh, what a nice cake!'

The cat's vocal repertoire seems without limits. Do you know the 'Hello! Anyone at home?' call? You hear this when you have inadvertently forgotten to unlock the cat flap, and your cat is demanding to be let in, first in a friendly, but later in an increasingly resolute tone of voice. I can only recommend that you react promptly, because on one such occasion one of my cats dismantled the entire cat flap without much ado! Or do you know the annoyed miaow that is sounded sooner or later during a brushing session, telling you in unambiguous terms: 'Leave me alone! I'm pretty enough.' I'm sure you could add to and embellish this list with endless examples!

One problem I get asked about relatively frequently is an increase in the really nerve-racking type of miaowing, sometimes in the middle of the night or at an unsociable early hour, or with cats who are getting on a bit. This is a problem that should be investigated further. Cats who are deaf can't hear themselves, and tend to miaow excessively loudly.

Well, to be honest, the banal miaow isn't really that important to us felines. But we have worked out that humans react to it rather well! When one of those **'tin-openers'** is talking to us, he or she will be really pleased when we **give them a miaow** every now and then. Well, a small price to pay in order to ensure domestic bliss! If the human is a bit reluctant or slow, or doesn't react at all, a long enduring miaow will make him more pliable, and in the end he'll be putty in your paws!

The senses

When cats get older, the performance of all their senses will decline, and this change sometimes makes them feel very insecure. You should be aware in this context how exceedingly well the cat's sense organs work under normal circumstances.

Their eyes only have the same visual acuity as a human's eyes. But in the half-light of dusk or dawn, cats can see a great deal more than we do. When the light is bright, a cat's pupils turn into long, oval slits. With reduced light, the pupils become rounder. The upper part of the background of the eye behind the retina is lined with special light - reflecting cells that act like a mirror. When a ray of light travels through the retina without being absorbed by it, the reflective layer will reflect it back on to the retina. This works like a light intensifier. The eye needs a tiny amount of light to do this; even a cat can't see in total darkness. If an elderly cat starts incessantly miaowing at night, a possible cause may be that the remaining light is no longer sufficient for him to see enough. Cats who have suffered a stroke are also sometimes affected. If a cat owner notices such changes in their pet, he or she should consult the vet.

In many cases it may be enough to leave a few weak lights on indoors overnight, such as children's night lights. These minimal additional light sources can make up for the cat's reduced ability to see, and make him feel less insecure.

Also very impressive is the hearing of a healthy mature cat: being an ambush hunter, the cat has large ears that can move independently to intercept sounds from several different sources at the same time. Cats are extremely sensitive to very high-pitched sounds; they can hear sounds at least one-and-a-half octaves higher than the highest sounds a human can hear. What this means in practice is that, for example, 1,000 mice would have to squeak all at the same time in order for we humans to hear the equivalent of what a cat hears when only one single mouse is squeaking.

When this remarkable ability to hear begins gradually to fail, it is not really surprising if an elderly cat gets frustrated and feels insecure. Unfortunately I can't offer any other advice than to consult the vet about it, who might prescribe medication that improves the circulation; in the worst-case scenario, the vet may temporarily prescribe anti-anxiety drugs, in order to help the elderly cat cope with the

Cat's eyes: perfect for hunting at dusk or dawn, but in complete darkness even a cat can't see a thing. (Photo: Schanz)

changes affecting his life and body. In any case, this would be a better solution than an elderly cat having to live a life of fear and insecurity.

If a cat miaows increasingly during the early hours of the morning and evening, this is often due to a desire to be let out. The time straight after sunrise and after sunset is the most interesting part of the day for cats, because this is when their potential prey is particularly active and therefore easy to pounce on. Because all cats are more or less subject to this biorhythm, most of their feline colleagues will be sneaking around the outside of the house at the same time. As much as you may find this early morning miaowing annoying, it is a perfectly normal behaviour. Therefore I can only give you one piece of advice: once your resolve weakens, and you give in to your cat in some way, just once, he will try it on again and again. As far as persistence is concerned cats can beat the pants off us every time!

Other cats literally learn to relate never-ending epics by miaowing incessantly at unsociable hours, because they know what sort of thing gets us going. Once a cat has got your attention, he has won. Whether this is negative attention (scolding, for example) or whether you climb out of bed in order to appease your moggy with some food is of secondary importance. This is what your cat will learn from the experience: miaowing gets your attention, and is therefore successful; and if it has worked once, it will work again! There is only one way to help cat owners who are terrorised in this way: they must be consistent and not pay any attention to this kind of behaviour.

The cat's large ears can turn independently towards different directions, and contribute to the cat's impressive sense of hearing. (Photo: Fotonatur.de/Askani)

Getting used to this new regime will prompt your cat to scream more than ever at the beginning. But the miaowing will gradually decrease. To be fair, I have to warn you that just when you think the cat's reconditioning has been successful (which will be the case after about two weeks), the cat will start miaowing once more, and with gusto. Behavioural therapists call this 'elimination'. Before a behaviour is eliminated from the cat's repertoire, it will reappear in an exaggerated form. Keep it up! You're nearly there!

Screaming

When we talk about screaming in the context of cats, we usually mean the beckoning, imploring, yowling, penetrating type of miaowing. A female cat on heat and ready to mate is more listless than normal, rolling around in front of the object of her desire – whether this is another cat or, by way of a substitute, her human – and she will be rubbing her cheeks on the ground while purring, cooing and screaming. The heat lasts between five and seven days. Even spayed cats can display this behaviour in a reduced form, because the spaying only prevents successful procreation, but has no effect on the relevant behavioural pattern.

At the end of the act of copulation, the female cat emits a defensive scream, similar to the way cats scream in a genuine emergency or during a serious fight.

When two tom cats fight, their yowling screams, almost reminiscent of singing, can be heard over long distances. These are not beckoning calls, designed to attract females, but the menacing chants of two males fighting for territory and resources.

No sign of being on heat, just pure relaxation. The rolling on the ground has a variety of communicative functions. (Photo: Schanz)

(Photo: Schanz)

Body language: facial expression and gestures

Facial expression is caused by visible movements on the surface of the face. The cat's face is very expressive because it has an extremely agile set of muscles in the area of the nose, mouth and ears. The whiskers underscore and emphasise every move of the muscles, and are equipped with very sensitive receptors. Through the ability to alter the size of the pupils rapidly, the cat's eye isn't just able to adapt to the changing light, but also to give clear communication signals to another individual as part of facial expression.

Have you ever closely observed your cat's face during play? If we humans had quicker reactions, we could say every time: 'Here it comes!' just a thousandth of a split second before the paw comes down. The cat observes, and very suddenly the whiskers fan out towards

the front, the pupils become perfectly round, all the facial muscles are tensed, and thwack! – the paw is right on target. Have a closer look – it really is fascinating!

Gestures, being the language of the body and its posture, can substitute, add to or underline the acoustic language. Facial expression and gestures are components of non-verbal communication, and are used in order to convey one's intentions and avoid misunderstandings. It is very important that you understand the following: even if a cat looks extremely menacing and uses all means available to him, in order to present a threat, his sole intention is to avoid conflict!

By contrast, a curious cat. It is easy to tell the positive tension from his posture. (Photo: Fotonatur.de/Meyer)

Phew, this is rather tiresome! Nothing but **boring theories**! Instead, let's have a look at the cat's superior ability to express his moods through body language.

Usually when cats fight, it is to a large degree a non-violent affair that doesn't cause any serious injuries to either of the combatants. Paradoxically, the reason for this lies in the very fact that cats are armed with such efficient weapons. Every attack harbours the risk of being harmed yourself. This initial fight avoidance through the use of a whole range of body language is called sparring. It is a ritualised fight; a precise, preordained and therefore for both sides largely predictable chain of behaviours is employed, in order to convey one's intentions and give an advance warning to the other party. If it hasn't been successful in resolving the conflict, sparring can of course develop into a serious, hot fight where real damage is done. This can happen when both cats involved are of about equal strength.

Attack, flight, aggressive as well as defensive behaviour, posturing, threatening, taking defensive actions, appeasing, fleeing, etc. – all these behaviours are summed up by the collective term 'agonistic behaviour' (Greek, agonistikos: possessing fighting spirit), and it applies to every conceivable conflict between rival members of the same species, which basically always involves the essential needs such as living space, food, procreation and raising offspring.

Neutral mood

A cat in a neutral mood displays a relaxed body posture, no matter whether he is sitting down, lying down or walking about. The eyes neither avoid eye contact, nor do they seek it. The ears are pricked up and react to sounds originating from the environment. The limbs are neither tucked under the body in a stressed unrelaxed manner, nor are they tense and ready to jump. The tail swings back and forth when the cat is walking around, and the fur is relaxed and close to the body.

Totally relaxed: a cat in a neutral mood looks at his environment in a laid-back fashion. (Photo: Fotonatur.de/Morsch)

Slightly annoyed: this cat is still undecided whether to avoid the disturbance or go on the defensive. (Photo: Fotonatur.de/Meyer)

Anger

An annoyed, angry cat shows his displeasure by initially beating and waving the tip of his tail, which can reach a crescendo of furious whipping. However, this tail whipping doesn't tell you anything about whether the cat will react in a defensive or in an offensive manner. On the contrary, it metaphorically shows the cat's actual mental state: the cat, together with his tail, vacillates between fight or flight, between running off in a huff and vehemently defending himself against the disturbance.

Defensive posture

The cat's defensive posture puts his whole body in a defensive mode – bearing in mind that the aim is always to avoid a serious physical fight. The defensive posture is in reaction to a threat, even if the threat is only a perceived one. The aim is to deceive by appearance: the cat does everything in his power to look larger than he really is.

In order to take things to the extreme, the tail can be made to stand erect and turned into a sort of a bottle brush with all the hairs standing

The little tiger is caught between defence and aggression, because his escape route is blocked. (Photo: Fotonatur.de/Schellhorn)

on end. The fur is puffed up, and often the cat assumes a slightly sideways position towards the opponent, presenting his broadside to him. The whole of this body image is called arching the back, and it is symbolic of what goes on inside the cat's mind. While the front end is already engaged in flight, the rear end is still bluffing. By doing so, the cat's body is foreshortened and the back is arched. The whole thing is accompanied by hissing at the opposing cat, or, if the opponent in question is a different species, such as a dog, by spitting. Small fake attacks are made, which only have one purpose: to stun and baffle the opponent, thereby creating an opportunity for escape. The cat will twitch forwards and backwards, as if an electric current was running through him. If he manages to confuse the opponent, it is possible for the rear end to become braver, and this gives the appearance of the front of the cat being overtaken by his rear end.

If this doesn't provide an escape route from the threatening stimulus, he will try to gain a strategic advantage by getting to a slightly elevated position. If this proves unsuccessful, too, he will remain on the spot, crouched down and motionless, anxious to protect the most vulnerable parts of his body. He can hiss and lash out with his front paws towards the enemy, or give a warning growl. The upper lip is pulled upwards, which creates small wrinkles on the nose, baring the teeth: 'Look what dangerous weapons I have, and you had better leave me alone!' Of course, this defensive posture can also turn into an aggressive one. There are no clear border lines.

By the way, hissing is generally followed by a blow with a paw, while an enduring, warning growl tends to announce a bite.

Fear

An acutely scared cat will pin back his ears so far that he almost appears earless. The extremely widened pupils are due to a strong influence of adrenaline, and point to a mental state of real fear or an extremely stressful situation. The whiskers are folded backwards. The cat makes himself small, hoping that the stimulus causing the fear will disappear. Often he'll sit with limbs folded under the body in order to protect vulnerable parts of the body, such as the neck and the tummy, and he'll try to squeeze himself tightly into a corner.

In the face of this tiger cub, you can see pure fear. (Photo: Fotonatur.de/Schellhorn)

Displacement activities

In conflict situations, cats often display displacement activities by preening themselves, or scratching or sniffing the ground. This happens when there are two conflicting but equally strong motivations that are blocking each other, prompting a third to manifest itself, in this context a seemingly displaced motivation.
This can involve being undecided whether to attack or flee, which is diffused with a displacement activity, such as the hectic licking of the fur. After inter - species conflicts, it can often be observed that the loser will diffuse his inner tension by demonstratively scratching some object (a tree trunk, for instance) – but only once the winner is out of sight. This letting off steam was called a gesture of spite by Paul Leyhausen.
The acute tension is diminished by the cat licking his nose and chops, and strenuously looking around him while avoiding eye contact with the originator of the conflict, and even more so if he is stared at or observed too rudely. This behaviour can be observed among big cats in the circus or the zoo, when – from the cats' perspective – they are exposed to the visitors' penetrating gazes.

This cat seems to have his ears pinned back with fear – but this time it's only due to a whim whilst playing. (Photo: Fotonatur.de/Meyer)

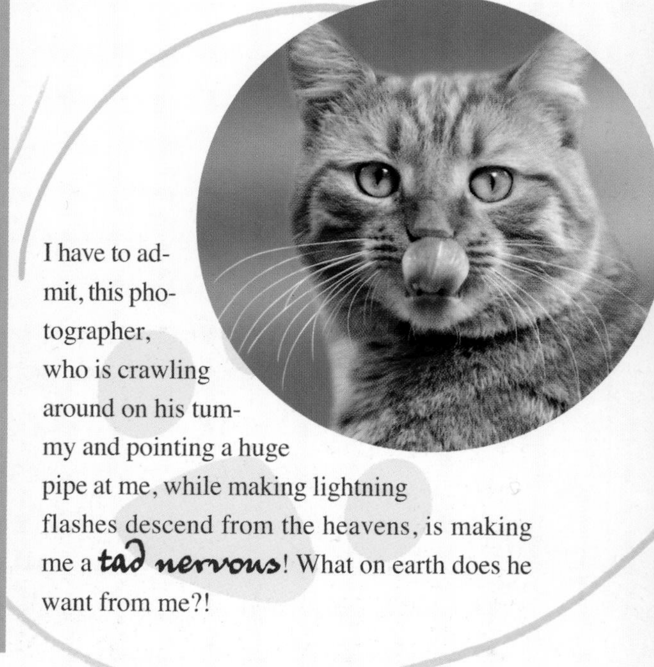

I have to admit, this photographer, who is crawling around on his tummy and pointing a huge pipe at me, while making lightning flashes descend from the heavens, is making me a **tad nervous**! What on earth does he want from me?!

All contact with the environment has consciously been broken off, and the facial expression seems to say: 'Just leave me alone!'
(Photo: Fotonatur.de/Schellhorn)

Depression

A cat can sink into a deep depression as a result of continual stress, which means permanent stress that is perceived as threatening (called distress), or from loneliness and/or social neglect. This kind of situation often arises when there are too many cats forced to live together in an inadequate space, or when one individual is constantly bullied by other cats. Animals that have lost a familiar partner – either a human or another cat – can also become depressed and grieve for months.

For an astute observer, the pronounced signs of this are easy to spot. Initially, the cat will neglect grooming his coat, the fur appears shaggy and loses its sheen. Then he will gradually turn away from his environment and appear introverted and defensive, showing no interest in anything going on around him. In the advanced stages, he will practically rest all day long. But this isn't relaxed sleep, it is the total cessation of contact with the environment. A clear sign of this is the fact that the cat keeps his eyes strenuously closed. He is usually lying in a closed off body posture with an arched back, the front legs folded tightly under the body, the ears in a sideways position, signalling rejection, while the head is tensely held up, but retracted (seemingly neckless). Of course, there is no cause for concern if a cat assumes this position every now and then, because may have suddenly felt disturbed by something whilst being stretched out for a relaxing sleep. If, however, he is sitting in this position often and for prolonged periods of time, and generally appears apathetic, it is high time to look for the causes. Being continually stressed like this will over time lower the cat's immune response, and lead to grave illnesses.

Typical of the attacking posture is the forward pointing position of the back of the ears. (Photo: Fotonatur.de/Askani)

Threatening behaviour

In contrast to defensive behaviour, a cat who is threatening or, rather, threatening to attack, is totally on the offensive, and the range of facial expressions contains all the nuances that can also be observed while the cat is eating. That's why threatening to attack is also known

The threatening posture is the opening gambit in the fight between uncastrated tom cats. (Photo: Fotonatur.de/Askani)

as threatening to bite. The whiskers are pointing forward and are fanned out widely, similar to a peacock's tail. The ears are folded sideways, but the root of the ear is pointing in a forward direction. This gives you a clue why tigers, for example, have white dots on the back of their ears: they only become visible to an opponent who is facing the tiger as part of a genuine attacking posture. Hence there can be no misunderstandings about the seriousness of the situation, because this visual signal clearly emphasises the intention to bite. Our domestic cat's ancestor, the genuine African wild cat, *Felis silvestris lybica*, has bright orange colouring on the back of its ears.

A cat can quite frequently and quickly end up in a situation where he has to display defensive behaviour. In contrast, the threatening attack posture occurs relatively rarely.

Tom cat fights

Uncastrated tom cats display this attacking posture during their fights to establish their position in the cat hierarchy. What is at stake here is power, status, access to potential mating partners and territory. A serious fight featuring everything from the cat's arsenal of attack, ranging from hissing and growling to scratching and biting, is preceded by sparring and ritualised threatening gestures.

To start with, the already familiar threatening posture can be observed: the tom cat makes himself large by arching his back and making his fur stand on end, he pulls back his lips, baring his teeth, and accompanies this with growling. But in contrast to the purely defensive posture, the opponent is held in firm eye contact, and stared at. The adversaries face each other in a sideways position, and approach each other on tip/toes – a strategy that makes the cat's whole body appear larger. As soon as they are nose to nose, and neither has given in, one of the tom cats will begin attacking the opponent's head and neck. In this situation, the objective is clearly to cause damage to the opponent, it's about all or nothing, either him or me, fighting for one's own interests. After the first of such attacks, the two tom cats will stare at each other once more, apparently with the objective of forcing the opponent into retreat. If neither of them concedes defeat, the second, damaging attack follows. This is followed by more ritualised threatening behaviour, and so forth.

By the way, tom cats have additional ways at their disposal for establishing a social hierarchy – the 'brotherhood'. This is exclusively a men's club, consisting of a number of uncastrated tom cats. It has a strictly hierarchical structure. The tom cat club travels around, jointly seeking out young, sexually mature tom cats in order to challenge them. The young tom cat has to subject himself to many heavy fights, before he is accepted into the club. These fights are exclusively about rank, not territory! The 'brotherhood' will wait in front of the young tom cat's house, beckoning and calling him, until he confronts the challenge. Once a tom cat has been accepted into the 'brotherhood' and has established his status, there are usually no more serious fights.

They start practising early: in this playful discussion, the kitten on the left is about to give in. (Photo: Fotonatur.de/Askani)

Cats don't do humility!

A gesture of submission, designed to appease the opponent or to make him more lenient, does not exist among cats. If a cat is losing a fight, he will be lying on the floor; but this posture does not mean submission, because he has all his weaponry at the ready, pointing upwards. This posture will not cause the superior opponent to end the fight. If the winner of the fight retreats, this is simply because he reckons that enough's enough.
We should remember this when we play with our cat. If a cat turns on his back and fights with his 'weapons' pointing in an upward direction, then this is no longer a game for him. He will feel defeated at this point. If on top of this we were to hold his tummy and cuddle it – as we humans often call it – for the cat it feels like he is being pinned down, as if we were saying to him: 'You are involved in a serious fight! I'm about to bite you.' It's no wonder that he will react in a fearful and aggressive manner, trying to hurt the human hand with his claws and teeth.
Rough fighting games are not as much fun for cats as some humans tend to believe. First, they scare the cat and put him in a position where he feels the need to defend himself; and second, instead of the human friend, we turn into this dangerous giant cat that needs to be fought off. As lovely as cuddling a fluffy tummy may be, it's only allowed if it's actually pleasant for the cat.

What do you mean, *humility*?
Explain. I can do curiosity, courage, fear, attack and defence. If I get into a scrap with my feline housemates, we may just pull out a few of each other's hairs, bite each other's legs, but not too hard – because that would actually hurt! Well, afterwards I just preen myself with gusto and – no harm done!
But I would definitely never, ever eat humble pie! Never! That would simply be too *undignified for a fine figure* of a tom cat like me.

The situation here is relaxed. The two cats know each other, the lying-down cat is laid back and does not feel defeated. (Photo: Schanz)

But let's get back to the massive conflicts between uncastrated tom cats (very rarely, castrated tom cats will be similarly aggressive). How intense and ready to engage in violent behaviour an (uncastrated) tom cat may be depends on his genetic and actual disposition, as well as on learned factors. The direct environment, i.e. the constellation of the local cat population in relation to the size of the territory, is certainly a big influence on a cat's readiness to engage in violent conflict. But what can you do, if you own such a fighter cat who will pick on and terrorise all the other cats in the neighbourhood, something that may affect your neighbourly relationships negatively and long term?

◆ **Castration:** The castration (surgical or hormonal) is a very small operation that will heal very quickly. The animal will not change its personality, and does not have to put on weight, if we humans take a little care not to overfeed him. On the other hand, the testosterone levels will go down a few weeks after the castration, and with it the cat's aggressive disposition. Your tom cat will not lose his cool as a result!

Sometimes the castration may only have been partially successful. The vet can establish via a blood test whether both testicles have in fact been removed, and there won't be a secret source of testosterone still hiding in the abdomen. I know of a tom cat that had to have five operations before the second testicle could be tracked down and removed!

◆ **Thyroid test:** Another possible cause of exaggerated irritability may be due to a malfunctioning thyroid gland.

◆ **Feeding:** You may also give a little thought to the feeding programme of your miniature Rambo. Some preservatives and additives influence the behaviour of our pets to an unknown degree. You can find out relatively easily whether this is affecting your cat, without having to resort to extensive lab tests Just give your cat only homemade meals for a few weeks, and observe whether this affects his behaviour. For more information on home cooking for cats, see cat magazines or other publications about the dietary needs of cats.

Also, never feed your cats in the garden. On one hand, by using the resource of food you will reinforce the territorial aggressive disposition in your own cat, i.e. the readiness to defend his territory against other cats. On the other hand, you will actually attract other cats with this food.

◆ **Controlled going-out times:** If your tom cat has been castrated and is physically healthy, but still remains a habitual thug, this can be due to the fact that he has had too many past successes, and his behavioural patterns are therefore already established. Sometimes this can be resolved with the help of a timeshare arrangement, worked out between you and the other cat owners. This timetable will curtail and regulate the outdoor time allocated to the little despot. It is important not to concede any of the important times to him; important and attractive times are mainly the periods around sunrise and sunset, because this is when prey is easy to come by. For real fighter cats, the late hours of the morning or the quiet afternoon period are best suited. These are the times when the cat's biorhythm makes him less agile. This is why the little thug has less of a chance of meeting his potential victims, and he will be able to carry out his patrolling duties with a lot more equanimity.

◆ **Working out post-traumatic shock:** Sometimes female and male cats undergo a personality change after an accident and become more aggressive. The physical and/or psychological shock leaves lesions on the brain that may trigger such a change. A single dose of the homeopathic remedy arnica is very useful for this, and should be part of every first aid kit.

Conflict situations between two cats, either unfamiliar or familiar animals, can lead to grave traumatic consequences. Cats learn incredibly quickly and have extraordinary powers of perception. As a consequence, two cats may have been getting on tremendously well with each other, or at least tolerated and accepted each other, when suddenly something fundamentally bad happens to destroy this delicate balance for a long time to come. This is particularly tragic if both cats are living in the same house! But there is no reason for the cat owner to despair, or to take the agonising decision to get rid of one of the cats. There are excellent ways to desensitise and countercondition in regard to this negative experience, in order to develop new and, above all, positive associations. For your sake, and your cat's, please seek advice and help from an experienced animal therapist. I won't pretend that the relevant therapy is going to be easy; it requires a lot of work and patience. But it is really worthwhile. It is relatively common for friendships between animals to develop as a result of such an individually tailored therapy, which would have been completely unthinkable beforehand. In such a situation, no stone should be left unturned.

A cat that has been reared by humans may have problems later socialising with other cats. (Photo Schanz)

Lack of socialisation

If a cat is aggressive towards other cats because of a lack of socialisation, the therapy will be a great deal more difficult. Cats live through a sensitive phase from about the 12th to the 14th week of their lives. During this time, the cat's brain has a window of opportunity for learning things that are important and vital for survival, lasting his entire lifespan. Things the cat has not come into contact with during this period will always be associated with insecurity for the rest of his life. If a bottle-fed kitten grows up without any contact with other cats, this experience is denied him, and as a result he will always feel uneasy and/or have problems when dealing with other cats.

Are you surprised that the social interaction among cats is not guided by instinct? Well, it is not always easy to distinguish whether a behaviour has been inherited, and is part of an ancient tribal knowledge, or whether it has been learned. Often we're dealing with a combination of knowledge that is part inherited and part learned. As a rule it can be said that the more essential a behaviour is for survival, the greater the certainty that it is innate and not learned. The greater the effort, time and energy spent on rearing offspring, the greater the amount that has to be learned, in various ways. The more instincts an animal has at the basis of its behaviour, the more broadly it will be able to develop its abilities, and the greater its ability to learn. But only the interaction with the inanimate and the animate environment guarantees a good development As a result, you get an animal with a large treasure trove of experiences, with the uninhibited readiness to engage actively with his environment, and last, but not least, the ability to encounter new situations, and to cope with them. I wish the awareness of this fact wasn't just commonplace in the context of the upbringing of dogs, but also with regard to cats! Our cats live in the same overstimulating environment as our dogs, and should therefore be familiarised with all the factors in a sensitive manner, and at the right time.

It makes my heart bleed whenever I hear that one of those cute kittens has no mother, or that he's been separated from her too soon. Don't they know that we like to stay with mummy until we're 12 weeks old?
Take it from me, raising kittens involves much more than just nursing them! We learn a great deal about life from our dear mama, just by **observing and imitating** her! Slowly, but surely we explore the known world from top to bottom. Our mothers, in their wisdom, know exactly when we're ready to learn our lessons, and they'll show us everything that matters!

Although the life history of my British short-hair tom cat Mogli is largely unknown to me, I can say with certainty that he was one of those kittens who were denied social contact with other cats during the decisive phase of his life. It is possible to get such cats to become used to other individual cats. Mogli now lives with me, and Anima and Sala. But every new cat that such an animal has to get to know means a fresh effort and a difficult learning process. In other words, one single positive experience is not enough to generalise a behaviour. What has been learned has only been learned with respect to the one cat that was the object of the learning process. If another cat is introduced, it has to be worked out from scratch once more that the same rules apply to the new cat, too. Of course, cats like this cannot be allowed to roam freely. They only understand cats that are familiar to them, and would therefore run into trouble as soon as they were to meet an unfamiliar cat.

At the beginning, I often squinted at Mogli in my efforts to teach him the basics of cat language. Among cats, squinting signals peaceful intentions, and it has a contagious effect. I thought that if Mogli were to squint at other cats, or reciprocate their squinting at him, he'd have an easier time of it. Well, Mogli is squinting. As if he were being paid for it! But he'll only squint at humans, not at other cats.

OK, that's an extreme case. Generally bottle-fed kittens aren't this antisocial towards other cats, because usually there would have been a mother or a fellow cat around from whom the kitten might have picked up a few things. But most veterinary practitioners will agree with me: there are no patients that are more hard work on the surgery table than hand-reared cats. During the decisive phase of their lives, they have experienced and got to know a human as their social partner, and they therefore relate to him not as a friend from another species, but as another cat. In extreme cases, they'll also use their claws on him, the whole works!

By squinting, cats signal their peaceful intentions towards each other. (Photo: Schanz)

Conditioned to a certain food?

Talking about sensitive phases in the life of a cat, I find myself confronted with the issue of food preference time and time again. Is there a conditioning effect regarding a certain food or not? I don't think so, because a conditioning to prefer a certain food would not make any sense biologically, and even be counterproductive. If it were true, then it would be absolutely impossible ever to find a new home for cats from a cat home, without knowing what type of food they had been fed on before; and because, unlike dogs, you shouldn't let cats fast for more than a day, they'd probably starve to death before anyone was able to find out what type of food they had been reared on – and owners of roaming cats would never have to worry about their cat free-loading at the neighbours'! Dr Mircea Pfleiderer, former pupil and respected successor of Paul Leyhausen, shares my opinion for the following reasons: African wild cats, the ancestors of our domestic cats, are forced to leave their mothers at the age of about nine months, and have to look for a territory of their own. In order to do this, they often have to wander into unknown areas, and this frequently means a change in the range of prey available to them. If they were, for example, exclusively conditioned to catch mice, they would starve to death in a different area where there may not be many mice, but instead other small animals, such as rabbits or rats.

Cats also learn from experience what is edible and what isn't. (Photo: Fotonatur.de/Rossen)

Cats learn from experience what is edible and what isn't. Many a young cat has to ingest a few shrews until he realises that these insect eaters are not digestible for a cat's stomach, and that this is the reason why he subsequently feels sick every time. The fact that a cat may have food preferences that he has been cultivating for years, only to change his mind completely from one day to another, has nothing to do with conditioning, but is a result of the idiosyncratic nature that we so much adore in cats; and, of course, the fact that we humans allow ourselves to be manipulated and wrapped around the little fingers of those cute cat's paws. I think it was Elke Heidenreich who once said: 'The cat is the only animal that succeeded in domesticating the human being.'

Idiopathic aggression

There is a particular, but fortunately rare, form of aggression that is called idiopathic aggression. Cats suffering from this real illness will attack humans, dogs and also other cats without any reason in a massive fit of aggression, and completely out of the blue. These attacks tend to be so extreme that they can cause really serious injuries. There is never a noticeable trigger. Affected cats are no longer themselves; from one minute to the next, there is no prior warning. Unfortunately, various methods of therapy, including changes in the way the cat is being kept and alternative therapies, have so far not led to the desired success. The origins of this type of aggression have never been discovered to this day.

Maternal aggression

Returning to our fighter tom cats, you may be wondering whether it is only tom cats that are aggressive and start fights, or whether female cats do, too. Yes, they do! But the classical aggressive stance can practically only be seen in adult female cats when they have to defend their young, or the territory necessary for rearing them. It is called maternal aggression. In other circumstances, female aggression stays mostly in the defensive range.

After the birth, the progesterone levels go down, and it is therefore assumed that this extreme readiness to be aggressive is dictated by hormones. Many owners who are lucky

When the mother cat behaves aggressively even towards her trusted humans after the birth of her kittens, this is due to post-natal hormones. (Photo: Schanz)

enough to witness a birth and the subsequent rearing of the kittens are shocked by the mother cat's extremely aggressive attitude, which is totally out of character. 'But she knows me, and must be aware that I'd never harm her or her kittens,' they complain. As a rule I'd recommend accepting the mother's behaviour, and leaving her and her litter alone for the time being, because things will go back to normal very quickly by themselves. If this aggressive behaviour exceeds the limits of what is bearable, a vet should be consulted, and preferably asked to come for a home visit. If things get out of hand, this natural protective mothering instinct can be influenced by the use of hormonal supplements. If the kittens are shielded by the mother from the domestic environment for too long, they won't be conditioned to get used to human contact.

Normally, the maternal aggression reaches its peak from the third week of the kittens' life onwards. This is the point where on one hand the kittens are independent enough to crawl out of the nest, but on the other hand are still a long way away from defending themselves, or being able to escape quickly enough in case of danger. In the natural environment during this period, the risk to the kittens is higher than ever. But at only five weeks old the kittens dart about with such agility, that the mother begins to relax and allows them a lot more freedom of movement. By the way, it is normal for the mother to get up after nursing, and lie down a small distance away from the nest, she rarely continues lying next to her kittens. This enables her to notice approaching potential enemies early and react more quickly.

The transport bite

The mother carries her kittens around by holding them with her teeth by the scruff of their neck. When she holds them in this manner the kittens become completely still, seemingly paralysed. This doesn't mean they are entirely

When the tom cat bites the female on the back of her neck during mating, he uses basically the same grip that the mother uses for transporting her kittens. (Photo: Schanz)

motionless. They just don't fidget about, they curl up a little and pull up their hind legs. In this position, the mother can easily carry them. As they're getting older and become too heavy and too large for their mother to carry, the mother cat will not flinch from resolutely grabbing any runaway kitten by one of his hind legs, and dragging him back into the nest.

The cat mother uses her canine teeth for the transport bite. She holds the kittens by the scruff of their neck. The transport bite and the mating bite that is used by the tom cat to hold the female by the scruff of her neck are basically carried out in the same manner. The only difference between the bite used for transport and mating and the bite used for killing is the biting inhibition. For the bite to kill, the canines are used for catching and killing the prey, and they bore through the victim's skin. During the mating bite, inexperienced tom cats can get it wrong at times, and accidents can happen when the tom is biting too hard – or, in other words, with too little inhibition – thereby injuring the female cat.

There is no clear border line between the transport bite and the bite to kill! Remember that when you hold your (adult) cat by the scruff of his neck – perhaps to prevent him moving about at the vets, or because you think you may be able to discipline your cat in this way. You would have to grip an adult cat quite hard in order to support his weight, so you'd only succeed in making him feel as if he was being subjected to an uninhibited bite. Reflecting on these underlying reasons, it suddenly becomes easy to understand why at this point many cats will put up a fight as if their lives depended on it. With respect to an uncooperative cat patient who turns every visit to the vet into a tortuous experience, you could try the following. Don't hold him down at all. For a change, just lightly lift the scruff of his neck, while slightly shaking him. It is surprising how well this method tends to work!

As far as disciplining cats is concerned, the gripping the scruff of the neck should also be avoided. We just don't have the ability to apply the right pressure the same way the mother cat does – and what good is a lap cat who will never again jump on the table, but neither will he jump on our lap, because we have destroyed his trust in us? What is right for kittens and cat mums isn't necessarily the right thing for cats and humans.

Goodness gracious me!

How uncouth is that?! Holding a cat by the scruff of his neck? I say! It's simply beyond the pale! All right, when I was a youngster, hormones raging an' all, I once tried it on with one of the other cats in the house, Anima. She duffed me up right and proper, like a tigress! It certainly wasn't one of my proudest moments …

Eye contact is an important factor in deciding whether a situation between two cats escalates or is defused. (Photo: Schanz)

Eye contact

You will surely have noticed from the previous descriptions that the distinction between attack and defence is determined to a large degree by eye contact. Staring at someone is provocative. Looking around, you can defuse a situation and avoid conflict. Squinting is contagious and signals peaceful intentions; it is how cats 'smile'. We should be very aware of this when dealing with our cats on a day-to-day basis. You should give it a try one of these days. If your cat is very busy doing something, watch him intensely and don't take your eyes off him. After a short while he will stop doing what he was doing, and in many cases walk away in a huff. Why? Because – from a cat's point of view – you were just too intrusive and terribly impolite with your persistent gaze.

We can use eye contact in a positive fashion by squinting at our cat or looking away when the cat shows clear signs of feeling stared at. You can soothe and calm an anxious cat by averting your gaze and looking around every time he looks at you. You can also stop your little tearaway from biting into your hand with unbridled gusto by directing a firm, compelling stare at him.

Have you ever wondered why your cat, who normally keeps well out of the way of strangers, will again and again, as if by magic, be attracted by those of your visitors who aren't particularly fond of cats? Next time, take a closer look at what's going. You will probably observe that the visitor concerned doesn't look at your cat once. In the eyes of the cat, it is exactly this that is making him such a polite and gracious human being – the cat just has to show him how much he appreciates his manners.

Further means of communication

Scratching and sharpening claws

Imagine you'd like to stop your cat from scratching the sofa. What are the means of communication available to you?

You may say: 'No!' in a strict tone of voice. The cat doesn't even look at you, and continues scratching! You lunge towards him and shoo him away. Your cat just looks at you as if to say: 'Are you quite mad to give me such a terrific fright?', and continues scratching. Or you may even grip him by the neck. This would be a total faux pas that might sooner or later cost you your cat's friendship. You attempt a hearty hissing noise, with the result that your moggy will just fall about laughing!

Now you have reached the point where you will reach for one of the cruder implements, such as the much-loved water pistol. Unfortunately, this disciplinary measure requires you to be present at the scene at the drop of a hat.

After a few days at the most, the suspicion is growing inside you that Kitty is scratching the nice sofa deliberately and with malicious intent – with the sole purpose of annoying the socks off you!

Let's leave our human perspective behind for a moment (never mind the cost of buying a new sofa) and, while he is ruining our furniture with casual innocence, look at our cat's broad communication spectrum. The cat hooks his front claws into the fabric, gives his body a thorough stretch, pins his ears backwards with an expression of great importance, and begins to scratch with power, endurance and an air of concentration. The claws of the two front paws are ploughing through the fabric in alternating movements. Apart from the fact that the pulling out and retracting of the claws is keeping the cat fit, giving his entire body a good stretching, he will at the same time get rid of the old worn out claw husks under which new sharp claws have already grown. The scratch marks left behind from his activity don't just provide clearly visible markings for humans and other cats to see, they also offer an olfactory signal for other cats, because of the small olfactory glands on the underside of the paws, which together with the sweat glands on the pads produce an individual, unique odour.

So while we were wringing our hands, anxiously fretting about our sofa being ripped to shreds with just a few strokes of a claw, the cat has performed a feat of perfect multi-tasking communication, laying claim to his rights, entitlements and status with:

- a visual signal represented by the whole body posture during scratching;
- the leaving behind of visual scratch marks;
- passing on an olfactory item of communication.

How well this method of inter-species communication works can be observed in households with several cats. Just after one cat has finished scratching, the other cats instantly feel the need to carry out the same activity in exactly the same spot. Scratching is obviously contagious. This way markings are confirmed, the other cats' 'perfume' is taken in and overlaid and mixed with one's own individual aroma, leading to a communal claim, an exchange of very individual markings. A social communication is taking place.

Some lone cats quickly realise that humans will react in a very similar fashion to cats. The cat only has to scratch the furniture and the human comes running towards him, giving him his full attention. Isn't this a successful interaction! The clever moggy will remember that this kind of behaviour will ensure the human's undivided attention and end his boredom in a thrice. It is of secondary importance whether you rush towards the scene of the crime whispering sweet terms of endearment or uttering wild curses. The fact is, the cat is the centre of attention.

While you'll probably find all this very illuminating, you may be far more interested in finding out what can actually be done in order to stop your cat from scratching the furniture.

The sharpening of claws isn't primarily designed for nail care, but for communication! (Photo: Schanz)

- **Create sufficient scratching opportunities for your cat.** These may be scratching posts, but also thick tree branches or an old wicker garden chair. Cats' preferences can vary greatly. It could be anything that will make your cat want to scratch it! It is important that the scratching implement is long enough to enable the cat to stretch its body to its full length. The material, too, is subject to preferences that may vary from cat to cat. The cat will actually show you which materials he prefers, by scratching different types of furniture. Equip these scratching zones and scratching posts with exactly those materials. One cat may prefer sisal or coconut matting, another one, soft, finely woven fabrics.
- **Choose a suitable site**. Cats will scratch most frequently near the place by the window where they like to sit and watch passing cats and birds. They also like to scratch near their sleeping place after waking from a nap. For best results, the scratching places should be positioned next to windows and next to the sleeping place respectively. A scratching post in the furthest corner of the room will rarely be used. Other strategically important passages and bottlenecks inside the home (entrance to the living room, to the kitchen, near the front door) are also places where cats like to scratch, so it would be a good idea to put something scratchable there. Scratching mats can be mounted on walls for this purpose. Some cats are visibly happy with just a piece of cardboard.
- **Encourage the cat to scratch on the surfaces provided by running your nails over them.** As we know, scratching is contagious and will stimulate the cat into joining in. Once his first own traces and odours have been introduced to the new scratching post, the cat will be eagerly refreshing them on a regular basis.

- **If you catch your cat scratching the furniture in spite of all these efforts, please don't punish him by shouting and screaming, shaking him by the neck or similar things.** A cat will learn very quickly that he can only scratch while your back is turned or in your absence. The often recommended water squirter is only of limited use. Your little moggy isn't stupid! He can easily see when you're holding a water squirter in your hand, and the cat knows exactly that it is you who is squirting him. More subtle methods are much more effective, and preserve the friendship between you and your cat!
- **Make the undesired scratching places unattractive.** You can, for example, apply easily removable double sided sticky tape or aluminium foil. Depending on the surface texture, you may have to install clear plastic sheeting first in order to protect the furniture.
- **Apply natural furniture care products that smell of citrus or eucalyptus**, which the cat will find unpleasant and will therefore avoid.
- **Put the precious tapestry armchair in a corner that is not deemed worthy of scratching in** (see above).
- **Some cats can be prevented from scratching by offering them food in exactly the preferred spot.**
- **Distracting the cat is useless.** Attempts to distract the cat by offering him food or playing with him may make the cat use the scratching action as a way of getting from you things he wants. Do play a lot with your cat. Kept busy and happy, he will not get ideas for creating mischief quite as easily: idle paws…

Choosing the right spot for the scratching post is also an important factor in deciding whether the cats will accept it. Near a window is ideal.
(Photo: Schanz)

- **Use eye contact for self-confident cats, but only those, please!** Eye contact is, as mentioned before, an important tool for cat communication, and by gazing into your cat's eyes with a fixed stare, you convey to him that this is your sofa, and that he has no business trying anything on here.

Installing cat-free zones

Some households have areas that the human inhabitants definitely and unequivocally declare to be cat-free zones – and why not? It is no different when the cat is outside: there too there may be another cat that vehemently defends his territory from being entered by another cat. Another garden cannot be traversed, because of a dog that has made chasing cats his hobby. But how does a cat learn that the human is claiming ownership of certain objects or areas? There is one basic rule: subtle methods are always better and more effective than any punishment.

Some cats can be prevented from jumping on your desk by the following method. If you catch your cat red-handed on the table, you can thoroughly sniff the exact spot where he had been sitting, and then cast him a severe and firm stare. Some cats aren't as easily impressed, but even for very obstinate specimens it is true that spoiling his fun is better than any punishment.

If, for example, the cat isn't allowed on the bed, you can spread a piece of polythene sheeting over it (builder's tarpaulin or something similar, cheaply available from any builder's merchant) and crisscrossing it with pieces of double-backed sticky tape. As soon as the cat jumps up on the bed, his little paws will get stuck to this disgusting stuff. With a little consistency and endurance he will quickly learn that the bed is an uncomfortable, unattractive resting place.

Big planters can be protected from cat action by using purpose-made cover grilles that are available from specialist pet shops. If the no-go area is a freshly made flower or vegetable bed, you can lay a trail of commercially available cat repellents, which you can purchase at any DIY and pet shop or online. Please make sure that the product you select is environmentally friendly and that it does not harm any other animals or insects. You can use tonic water to protect sandpits.

As much as most cats hate it, a closed door is sometimes a good way of making the territory inside the house interesting.
(Photo: Schanz)

My experience indicates that prohibitions act like a magnet to a cat. Everything that is strictly forbidden becomes a sort of a hobby and will be pursued again and again, sometimes demonstratively under the very eyes of you, the owner, other times secretly while you are not there. Make sure that you don't fall for this game of cat poker. You can deter your cat from jumping on the dinner table by simply carrying out every unpleasant procedure known to the cat on it, such as applying flea treatment, brushing, cutting claws, tick removal or administering tablets. This particular location will quickly lose its charm.

As a rule, every disciplinary measure requires you to have more stamina than your cat. One or two days isn't enough. How long it will take until the forbidden place will be avoided on a long-term basis depends on the consistency displayed by the humans – and, of course, on how persistent your cat is. Patience will pay off in the end! Even if punishments get you to your goal quicker, you use them at the expense of the trust between human and cat. Sometimes it may only be in the short term, but unfortunately it can also last a long time.

Secretion from the anal glands

Another means for conveying messages that cats have at their disposal is the use of a secretion from their anal glands. This secretion cannot be expressed voluntarily. When the cat tenses his muscles on the anus, i.e. when he defecates, the contents of the gland should normally be discharged. The secretion consists of a fatty liquid and of dead cells. For the human nose the smell is – to put it mildly – unpleasant. The cat uses it for individual marking.

The anal glands can become blocked and unable to discharge their contents by themselves, which often leads to the cat constantly licking his bottom or sliding around on the ground in a seated position. In this case, the anal gland should be squeezed out manually by the vet. This is not a very pleasant procedure, but the cat will feel visibly more at ease afterwards.

If several cats live in the same household, and relations between them are not at their best, the cat that is lowest in the hierarchy can smell more intensely of this secretion. In his subconscious, he obviously doesn't have the self-confidence to discharge the secretion and thereby add his tag to his faeces. Persistent diarrhoea can also lead to a blocking of the anal gland and cause inflammation. Some cats react by becoming unreliable in their toilet habits.

This matter ought not to be confused with stud tail, which occurs almost exclusively in uncastrated tom cats; castrated animals are rarely affected. It is caused by the secretion of a greasy substance from the sebaceous glands positioned at the root of the tail. If the site gets inflamed, it makes the cat feel uncomfortable. Stud tail can be treated with a powder or, in the worst case, by washing it. Exhibition animals will have points deducted for stud tail.

Rubbing and cuddling up

But the cat still has far more ways for leaving messages at his disposal, which are imperceptible to the human nose. He obtains his own personal body odour from the sebaceous glands on his jowls, his chops, on his chin and on his back just in front of the root of the tail, and this is transferred on to objects, other cats or on to humans, by rubbing or cuddling against them. Cats who are familiar with each other use a mutual gentle nose-butt and touching or rubbing of heads by way of a greeting, or, for a more formal occasion, sniffing the other cat's fur or anus.

I'm rather a shoe fetishist myself. It's not just the smell of my human, but also the **smells** that are carried into the house from the outside that make shoes so **exciting for me**! I roll around and rub myself up against them, adding my particular aroma, and this creates a delightful mixture that says to me: 'We all belong together! This is my home.'

The sebaceous glands on the various sites of the body ensure that when your cat goes round your legs, he is marking his own human with his odour. (Photo: Schanz)

The cat has a highly developed sense of smell that provides him with a range of information, even if the messenger is long gone. (Photo: Schanz)

Sense of smell and taste

The cat's sense of smell is important for the assessment of food quality. Otherwise, its main and most important function is in the sexual - social field. The smell markings with their combination of visible and olfactory signals are one of nature's more useful and sensible designs. An onlooker can work out what the cat is saying from his special body language signals. In addition, the chemical odours left behind last a long time, so cats can also convey a message to those who don't happen to be watching. This is nature's way of ensuring that, on the one hand, a potential social partner can be addressed directly and unequivocally, while on the other hand, the olfactory residue will last for a while, like a classified advertisement.

When a cat is intensely sniffing or licking a particular spot, he has located information about another cat and is busy analysing it. Like many mammals, he has a special organ inside the roof of the mouth, the Jacobson organ, which is designed to perceive sexual odours. In order to intensify the smell and to absorb and evaluate it with the Jacobson organ, the cat pulls back the corners of his mouth while the mouth is open. At this moment, the cat is all concentration, but to an observer it might look as if he were grinning.

Rolling and wallowing

By rolling around on the floor, the cat is expressing a sense of well-being. At the same time, he marks the location with his odour and in turn takes on the odour of the location. Cats will also roll on their backs during relaxed play and dangle after a toy in an upside-down position.

The rolling around displayed by female cats that are ready to mate tends to look a little more striking. Whilst in heat, she is more listless than normal, and rolls and wallows in front of the object of her desire, which may be another cat or, by way of substitution, a human being. She will rub her cheeks on the ground, purr, coo and plaintively scream her desire for all the world to hear. Female cats are in heat twice a year for about five to seven days. Female cats that have been neutered can still display the signs in a diminished form. The spay operation merely prevents the ability to procreate, but doesn't change the behavioural patterns. During the heat, the chosen social partner is addressed on three channels of communication simultaneously. The first is the visual channel, by displaying a clear sequence of movements; this is often underlined by sounds, via the acoustic channel; and in addition olfactory signals (smells) are left behind. These plucky loners don't leave anything to chance!

Rolling around on the floor is a sign of well-being and relaxation, and the cat is also leaving her odour on the ground. (Photo: Schanz)

(Photo: Fotonatur.de/Meyer)

Social structures

Why all this paraphernalia for communication in the field of social interaction, when the cat is well known for being a complete loner anyway? It is particularly important for animals that don't live in groups, packs or herds that establishing contact between different members of the species happens without risk and without misunderstandings, because a message that is misunderstood may result in grave physical injuries. But what about our cats? Are they really as antisocial as is often alleged? Actually, no mammal can possibly be antisocial, because, as we all know, mammals nurse their offspring. The rearing of their young involves quite a considerable effort! If an animal really was antisocial, caring for its offspring in this manner would be impossible.

The often observed solitary life style of the cat has very little to do with the social background, and rather a lot to do with, for exam-

Despite – or perhaps because of – the cat being a pronounced loner, social communication has a very important role to play. (Photo: Schanz)

ple, feeding and hunting strategies. Cats living in the wild primarily feed on small mammals. It would be counterproductive to hunt a mouse in groups of several animals, especially since one mouse is not enough to feed several cats. That's why cats hunt alone; and because hunting takes up the greater part of their lives, the picture that presents itself to the onlooker is one of a solitary animal. But first impressions are often wrong. Let's have a closer look at the hidden social life of the cat.

Cats are actually very social beings. Perhaps the reason why this often goes unrecognised is because they have such a large variety of different subtle social structures.

The observation of domestic cats has shown that the cat is enormously adaptable. Tom cats living on a farm, who are not directly subject to human influence, tend to leave and go out into the world as soon as they have reached sexual maturity. Adolescent females often stay in the mother's territory and live in groups. In large feral cat colonies, which can be found, for example, on archaeological sites or industrial estates, there is a hierarchy of small, matriarchal groups.

Favourite spots in one's territory are visited frequently. When doing so, encounters with other cats are avoided. (Photo: Schanz)

Territory of outdoor cats

The classic territory of a domestic cat looks like this. A tom cat's territory is about three times the size of that of a female cat, and it can contain one or more female territories within it. The average size of the territory of a cat living in a rural setting is about half to one square kilometre. Female cats defend their territory much more vehemently than tom cats, and the borders are drawn much more precisely. Due to the laws of nature, they have more to lose, because they need a secure territory for raising their young. This is conditionally also true for neutered females.

Within the territory, there is a primary residence representing the nucleus, which contains at least one secure sleeping spot and a potential place for raising the young. This may be the whole house or flat, but it can also be just a single room or a quiet, secluded corner. The rest of the territory is called the roaming area, which is crisscrossed by a cleverly devised network of paths. These trodden paths lead to the most attractive places within this area: a field, where one can go hunting, or a compost heap where lots of plump mice are just waiting to be caught. However, this can also be a preferred observation post or a sunny spot for lazing around.

The network of paths that leads to all these locations within the roaming territory is regularly patrolled. Other cats also use it, but not at the same time. Encounters are carefully avoided, and the unwritten law applies: whoever is there first has the right of way! Where there is no clear field of vision, there is always the chance of an unexpected encounter, and even outbreaks of hostility. But these relate only to this particular location and this particular time. In a different location and at a different time the underdog can actually come out on top, because the closer he is to the centre of his territory, the stronger and more self-confident he will become. This ensures that a cat cannot be chased off his territory by another cat.

A social gathering

Cats living on farms, as well as in domestic households, are known to meet up occasionally and just hang around each other for a while. Paul Leyhausen called this social gathering. Female and tom cats set out, cross the borders of their territories – notwithstanding whether it happens to be the mating season – and meet in a neutral location. They just sit there, often for hours, in relatively close proximity to each other, whilst keeping to the required individual distance. The mood is calm, peaceful. There is no hierarchy at these meetings. Afterwards, each cat goes its own way once more. Female and male cats return to their territories and their real lives. It is possible that during this social gathering two tom cats may have been lying peacefully next to each other, who normally can't stand the sight of one another. Why? Nobody knows.

I live here with two other cats, but one thing is for sure: **the sofa is my personal space**! It's not just very comfy, but you also get a very nice view of all the toys on the floor and into the other rooms. This is not negotiable. I will not share this sofa with any other cat!

Territory of indoor cats

The territorial hierarchy, subdivided into time and space, often has great importance attributed to it in regard to behavioural therapy for indoor cats. Usually the households concerned consist of several cats, with one cat (or several) starting to mark his territory or forgetting his toilet training, or there is the sudden outbreak of a cat war. Meanwhile, the owners are wondering: what has happened here?

Just sitting together for a little while: the precise meaning of these cat meetings has not yet been established. (Photo: Fotonatur.de/Meyer)

In the case of free-roaming cats, the primary residence may be, as mentioned before, the house or the flat, or perhaps just a single room. If you allow pure indoor cats – obviously with the best of intentions! – the unrestricted freedom of your whole home, in terms of space and time, they possess a large primary residence, but no roaming area. All the rooms are used equally by all the cats and at any time. In the long run, this situation can become the cause for conflict.

In such cases, the therapy will look at partitioning the living space and the time when it can be used. Or in other words, we attempt to create an artificial roaming area. Unfortunately, it is impossible to set up a fixed therapy plan here, because every home has a different design, every group of cats has a different dynamic, and last, but not least, the whole plan has to be adapted to the owner's life style. You can take this as a rule of thumb, however: free space and an interesting, stimulating environment for the cats – yes! But you should have a critical look at your home and how it's being used by the cats. Are there individuals who are barred from entering certain rooms by other cats? Are there rooms that are favoured by individual cats? Would it be possible to close off individual rooms, and to allow access to them for, say, only a few hours each day? Being allowed to roam here would be the icing on the cake, the highlight of the day. By using your powers of observation, you will get a more profound insight into the relationships the cats have with each other, and problems with toilet training can be resolved or avoided altogether.

Territory when moving house

Some owners of free-roaming cats may ask themselves whether, if the partitioning of territory is such a complex affair, they should actually take the cat with them at all when they move house. My answer is: yes, in spite of everything. Of course, it will be necessary for the cat to familiarise himself with the new place, and this generally means four to eight weeks' house arrest. Each house, each environment is unique – not just the way it looks, but also how it smells and sounds. First of all, everything has to be explored. But cats that live in a close relationship with we humans are unhappy if the person to whom they're closest suddenly disappears. They have to find new human attachments, and this isn't any less of an upheaval than moving to a new home!

Of course, cats are not nomadic and are therefore attached to the place where they live. But normally they have a stronger bond with their humans than with a house or a flat – at least, on behalf of every owner, this is what I hope to be the case! Once at the new place, it is not advisable to release the cat into the great outdoors too soon. Increased miaowing, especially in the first two weeks after the move, is rather a sign of insecurity than the urge to run around outside. In spite of the many astonishing reports about cats that have found their way back home after many months and even more kilometres, at present we have no knowledge of a special homing ability. What is known, on the other hand, is the fact that the likelihood of a cat finding his way back home

Moving house is always stressful for a cat, but with a few simple tricks you can help him through these difficult times. (Photo: Schanz)

is drastically reduced when the distance exceeds five kilometres.

While the cat is confined to the house, you can put this time to good use by doing some 'finding your way home' training. For this we will take advantage of the good old conditioning reflex, which was researched by the medic Ivan Petrovich Pavlov. Whenever the cat gets his meal, a special sound is made half a second before the food is given to the cat. This can be the sound of a bell, a whistle or, if you are very regular in your habits, you could even use the sound of a church clock or a grandfather clock chiming on the hour. If done with consistency, within a few days the chiming of the bell – or whichever sound – will suffice. As soon as the cat hears 'his' sound, he will run towards the kitchen like greased lightning in expectation of the food! When your moggy is able to enjoy his freedom once again, this pre-training will make him appear precisely on time on 90 out of 100 days. On the remaining 10 days, the cat will be busy doing something more important – and we humans will just have to live with that.

Territory and toilet training

The subject of toilet training is very important to many cat owners. Not every 'business' that misses its intended target necessarily means that a territorial dispute lies at the heart of the matter. Let's take it from the beginning: how is a young cat toilet trained?

In principle this is very easy. As soon as the kitten shows signs of needing to go, you just pick him up and put him into the cat toilet. If, on occasion, he misses, you just transfer the 'business' into the toilet to help the little fellow find his way the next time. Please, whatever you do, no scolding or, even worse, rubbing his sensitive little nose in it. This way the kitten will only learn that 'having to go' is a very bad thing indeed, and as a result he will look for a secret, secluded spot for his next business. If you notice that the little kitten urgently needs to relieve himself, but is still a little undecided about where to go, you can just scratch the cat litter granules with your fingers a little. The kitten is all curiosity and will instantly hurry towards the thing that is making this entertaining scratching noise. And, bingo! he is right where he wanted to be.

If a cat doesn't use his toilet the way he's supposed to, you should think about the number of cat toilets available. Every cat needs two toilets, because in the great outdoors cats don't like to urinate and defecate in the same spot.

There are many possible reasons why a cat has problems using the cat toilet, ranging from taking a dislike to the litter granules to the cat litter being too close to a favourite sleeping place. (Photo: Schanz)

For each additional cat, you have to add another toilet. That means two cats need three toilets, three cats four and so on. It will be worth the effort! Please don't think that this cannot possibly be the reason. My cats have been together for several years and one toilet has always been enough for them. This can work out well for many years, and all of a sudden, it's no longer sufficient. The cat toilet is like a communal laundry room in a block of flats: a source of potential conflict.

The cat litter has to be given special attention, because there are cats that develop certain likes and dislikes. You should try out several different kinds of cat litter. It would be great if cat litter manufacturers were to offer trial packs, just as cat food manufacturers do.

If a cat suddenly forgets his toilet training, this can be due to a negative experience during a visit to the toilet. Maybe something startled him, or he has a bladder infection, making every visit to the toilet a painful affair. Offer him a different toilet, possibly in a different place and with a different type of cat litter.

If a cat has never learned how to use a toilet properly, or has forgotten it during a stay in a cat home, you can make things easier for him at the beginning by putting a layer of potting compost or sand on top. Gradually reduce the amount of potting compost, until the cat knows and accepts the cat litter.

Closed and open-top cat toilets are available. The closed top models can be equipped with a charcoal filter, which is more for the benefit of the humans than for the benefit of the cat. Some cats will have no problem with accepting such toilets; others, on the other hand, will.

The position of the cat toilet is also very important. If there are water or food bowls or, moreover, a preferred sleeping place nearby, the cat will avoid using this toilet. Cats are clean animals, and to them – as it is for us – it is not acceptable to do their business in the same place where they have their meals or like to take a nap.

Urine marking always happens for a reason!

The problem of urine marking has to be distinguished from the problem with toilet training. The difference is not how much urine is deposited, but exclusively where it is deposited. Cats with impaired toilet habits always deposit their urine horizontally, while urine marking always leaves a vertical mark. Female cats as well as castrated tom cats are capable of spraying. This can just be a few drops, but can possibly involve quite substantial squirts. The 'record' in spraying performance, as far as I know, is held by the uncastrated tom cat belonging to a client of mine, who managed to spray into her face while she was standing behind him! I do hope that you will suppress your laughter, and feel a little compassion for this lady!
As the name suggests, this spraying is about marking territory and the objects contained within.

This can be the laundry basket, full of clean or dirty clothes, the coffee machine or you, the tin-opening slave. Once you get inside a cat's head, it quickly becomes clear that when a cat feels safe and protected inside his territory, he doesn't need to constantly keep marking his environment, and in the worst case this could be on a daily basis. Something's up! What could be making a cat feel so insecure that he feels the need constantly to renew the claim to his property? Is it the 'discussions' with other cats living in the same household? Or is your spray artist an indoor cat who is becoming extremely agitated because of a rival who is making an exhibition of himself below the kitchen window? Or is your cat marking his territory because the newly installed cat flap is making him feel insecure? Or did you perhaps spend an unusually small amount of time with your moggy in recent weeks, and he has understood that, if nothing else helps, a scented reminder will attract your full attention? In some cases, the search for the why and wherefore is pure detective work, but there is always a reason. Please ask for expert advice.

Out of doors, urine marking is not a problem, but inside the house it is a sign of great insecurity. (Photo: Schanz)

Cats can develop intense friendships if their personalities match. (Photo: Fotonatur.de/Meyer)

Who will get on with whom?

If you are thinking about whether you should get a second cat to go with your single cat, there'll always be one obvious question: who will get on with whom? There are, of course, cats that prefer to live as 'only' cats with their humans. Simply to substitute a deceased playmate with another cat is often also not advisable. If there was a close bond between the deceased and the remaining cat, this does not automatically mean that a new cat will be welcomed in the same way.

The best combination is probably to keep a pair (of siblings) together who are already getting on well, and to take in both cats. If you already have an adult cat, the best chance of him accepting a newcomer would probably be if the new cat were young, at the most two years old. The age difference should not be too great, however, in order to avoid the old cat becoming too stressed and annoyed by a young nipper.

Getting a female and a male cat to get used to each other is probably the easiest option. However, tom cats are able to forge intensive friendships with each other, as can two females. You just have to be aware that each gender has its particular peculiarities. Female cats are by far the more territorial. For them, it's all about the territory. For tom cats, it's more about social hierarchy. Castrated tom cats react in a more 'female' manner. Leaving male kittens with their mother seems to me to be the most problematic constellation. Once a kitten is about nine months old, it is very possible that the mother will reach the conclusion that the time has come to chase off the adolescent cat so that he may become independent and find his own territory. After all, according to nature this is the normal course of events.

But even more important than age or gender is whether the personalities of the two cats are matching. If you already have an extremely affectionate lap cat, you should not take in another animal of a similar disposition. Problems with cats competing for your affections would be inevitable. A little rascal, full of energy, is best matched with a young cat that is up to this level of activity, so both can benefit from playing together. An introverted cat may benefit from a cat who prefers the company of another cat to bonding with human beings, in order to prevent the introverted cat from sulkily retreating into a corner altogether. Whatever combination you are aiming for, I can only advise that you ask the breeder or the cat home for a trial period. If it turns out that the newcomer isn't to the old cat's liking at all, the whole transaction can be reversed. Friendships between cats cannot be enforced and they are purely based on tolerance.

Play and matters of consequence

I get really passionate about playing, once I manage to forget that the mouse isn't actually real. With a little imagination, I'm even able to overlook the fact that the bird that was once attached to my crow's feather is no longer there! Then I work myself up into a proper hunting frenzy! My personal favourite is fishing for prey under the carpet. You can pretend you're poking around inside a mouse hole. Great fun! Sometimes I have to stuff the toy mouse under the carpet myself (you just can't get the staff these days). Other times my human does join in by poking a stick under the carpet, and I get to attack the resulting bumps.

Stalking is an innate skill, but nonetheless training is necessary in order to hone and perfect it. (Photo: Schanz)

Why cats play

For highly developed mammals, play has the function of training and employing the same movements and behaviours – simply the whole behavioural repertoire – in the same way as in a real life situation. Play is, so to speak, about training for real-life situations, and limits are being tested, one's own as well as those of one's playing partners.

The motor skills necessary for catching prey, such as stalking, ambushing, grasping and carrying, are not learned in the conventional sense, but rather they are matured. These segments that make up the hunting activity are innate. The only thing that has to be practised during the maturation process is how to employ these actions adequately, confidently and above all successfully. Practice also strengthens muscles, bones and the cardiovascular system.

All the segments that make up the hunting activity mature gradually. In our domestic cats, you can recognise a sequence of events that is usually, but not necessarily, kept to. First, the kitten trains all the actions that are involved with stalking and grasping the prey. This is followed by working on the areas that require refinement, such as stalking and ambushing. Only at the very end does the act of killing mature. This sequence of events is well founded. When kittens grow up with their siblings, as is normally the case, they practise all these actions on each other. To make sure that nobody gets seriously hurt, the act of killing only matures right at the end. As the result of domestication and the stagnation at the adolescent stage, many domesticated cats only retain an incomplete (if any) ability to carry out the act of killing. There are two reasons why a cat brings a mouse into the house but doesn't kill it: either the cat is scared of the mouse, perhaps because he has had a bad experience and has been pinched on the nose once before; or the act of killing has not reached maturation, and the cat is acting like a kitten. This may never change. Whatever happens, no matter what our cats bring home, whether it's dead or alive, furry, slimy or feathered, we should always praise our cat – and at least pretend that we are extremely pleased. From the cat's perspective, these are gifts for his beloved human!

In order that both parties can enjoy the game, the human should know his cat's preferences. (Photo: Schanz)

Play is meant to be fun!

How can you tell whether cats are just playing with each other rather than fighting for real? The determining aspect of play is the absence of any relevance; nothing is at stake. Another decisive aspect is the fact that during play there is a constant switching of roles, for example from hunter to hunted, and behaviours from all sorts of functional areas are mixed up together in ever varying combinations. However, if as a cat owner you observe that it is always the same cat that is the hunter and the other cat that is always the victim, then you can safely assume that this is a real dispute and not a game.

Prick up your ears, stalk and **attack**!
That's our world! OK, grown-up cats think before they pounce. They evaluate the situation regarding its potential for success. I'm a young red-blooded tom cat, and to me this sort of thing seems totally boring!
Why not just go for it? Who dares, wins!
And practice makes perfect.

Regarding playing interaction between human and cat, I often hear complaints such as: 'My cat doesn't play. It always just about chasing and biting.' Other cat owners are saying: 'My cat just watches when I want to play with him, and doesn't join in at all.' Such behaviour can have several reasons. Cats are stalk and ambush hunters. They crouch with all their muscles tensed and watch what the potential prey (or the toy) is doing, how it's behaving, and when the right moment has come to pounce. Watching is a normal component of the game or, for that matter, the hunt. Some humans just lose patience too quickly, and end the game before the cat has begun to participate.

It is also possible, however, that in the cat's eyes the toy may be too large and seemingly too dangerous. If this is the case, he will just sit there and suspiciously observe the situation. The fear of a toy is characterised by the cat looking around a lot. He is averting his gaze in order not to cause a provocation, and to defuse the situation.

Another reason for a healthy cat not to join in the game is often the way humans play. In order to initiate a game that the cat can truly enjoy, we have to know which natural triggers will set off the cat's hunting behaviour. After all, the toy mouse or a bunch of feathers is supposed to represent the prey, which is to be played for. First of all, it is scratching and rustling noises that initially draw the cat's attention to his 'fellow creatures', and which make him begin the search. In order to activate the instincts of stalking, ambush and pursuit, something has to move rather quickly on the ground. It must not be too large and should move away from the cat in a sideways direction or straight ahead of him. This is because real prey flees, hides and runs away in zigzag movements, as it tries to escape. Prey doesn't run into the arms of the hunter and jump up and down on him in an annoying fashion. I observe again and again how some cat owners of bother their cat with the toy, and almost torment him. It's a pity, because for every cat playing is most enjoyable when the human is involved.

Especially for cats that are kept indoors, playing is immensely important for their physical and mental health. Owners should take the time for play on a daily basis – even if they have

The cat is and will always be a predator, but he tends to catch mostly weak or sick birds. (Photo: Fotonatur.de/Morsch)

several cats! There is nothing wrong with the playing action verging a bit on the wild side. If the cat is tired afterwards and collapses on his side like Pinocchio with his strings cut, this is not a problem at all, as long as the cat isn't scared or stressed. For an indoor cat that doesn't have the opportunity to stalk mice or birds, there is nothing better than to be well and truly exhausted once a day.

Attention: potential risk from cat toys!

Allowing a cat to play with a piece of string, cord or a shoe lace on his own can be very dangerous. Due to the texture of the cat's tongue he cannot spit out objects that have a rough surface texture. To avoid the risk of choking from swallowing very small prey in one piece, he will check the direction of the fur or feathers by running the hairs on his chops over the prey. The tongue is equipped with horny papillae. They facilitate a quick swallowing of the prey. Any rough material, once picked up by the tongue, cannot be spat out again by the cat. Instead it works its way towards the throat and can cause choking and knotting of the bowel. A piece of string made from leather is much better suited for fun and risk-free playing.

Is the cat a bird-killer?

Cats are predators, and whenever the conversation turns to the consumption of prey, we will soon arrive at an old and evocative subject. Are cats bird-killers? Of course, every cat will catch the occasional bird during the course of his outdoor life. However, it is mostly old, weak or sick birds or very young nesting birds that tend to fall victim to the cat's birding efforts.

As mentioned already, the stimulus that triggers a cat to become aware of potential prey is a scratching or a rustling sound, which is made by birds, mice and squirrels alike. The cat hears this sound, pricks up his ears, locates the source and stalks and ambushes the cause of the sound. This is often observed by irate bird lovers who will intervene in order to chase away the evil hunter. It's a pity, really. Upon further observation it would become obvious that the first impression was wrong. After locating the prey the cat hunts by sight. He observes the prey and waits for a suitable moment in order to grasp it. But song birds especially are very lively, quick and temperamental. After a very short period of time the cat, being a sight and ambush hunter, becomes tired of observing these hyperactive birds. Everyone who has ever taken the time to watch this game will be able to confirm that most cats will saunter off upon being pursued by screeching, irate blackbird or robin parents, rather unnerved by the experience.

Final words from Sala

Hey you, the human over there!
We cats are unique.
Have you actually got that now?
We allow you to experience a world that is not your own. Cats are mystique, dedicated love and reserve all rolled into one. When a human bonds with a dog, over the years their relationship begins to resemble married life. Both parties are sure of each other. The first flush of love turns into harmony, closeness and a bond of quiet mutual love and trust. You won't get that sort of thing with us cats. We will always be the love affair of your life. We will never be completely yours. The lure of the unattainable will always surround us. We have mastered to perfection the art of making our exits just at the moment when things are at their most beautiful.

We cats are torn between our overwhelming need to be free, our individuality and our great desire to share our lives with you humans.
We don't do everything the way you'd like us to. But everything we do, we do of our own free will, and with complete intensity and dedication. Human beings who help us cats with every new day to complete this tricky tightrope walk between forming a close bond and maintaining our individuality can be assured of our loyalty.

All the best to you, friendly tin-opener,
and to your feline friends.

Yours truly,
Sala

Further reading

Henry R. Askew:
Treatment of Behaviour Problems in Dogs and Cats: A Guide for the Small Animal Veterinarian,
Wiley Blackwell, 1996

Martina Braun:
Clicker Training for Clever Cats – Learning can be Fun!
Cadmos, 2009

Paul Leyhausen:
Cat Behaviour: Predatory and Social Behaviour of Domestic and Wild Cats,
Garland STPM P, 1979